Read for a Better World™

EARTH

A First Look

PERCY LEED

Lerner Publications ◆ Minneapolis

Educator Toolbox

Reading books is a great way for kids to express what they're interested in. Before reading this title, ask the reader these questions:

What do you think this book is about? Look at the cover for clues.

What do you already know about Earth?

What do you want to learn about Earth?

Let's Read Together

Encourage the reader to use the pictures to understand the text.

Point out when the reader successfully sounds out a word.

Praise the reader for recognizing sight words such as *the* and *are*.

TABLE OF CONTENTS

Earth

This is Earth.
Earth is part of
our solar system.

Earth
the sun

Our solar system has eight planets and the sun. Earth is the third planet from the sun.

It takes Earth one year to move around the sun. As it moves, Earth turns like a top.

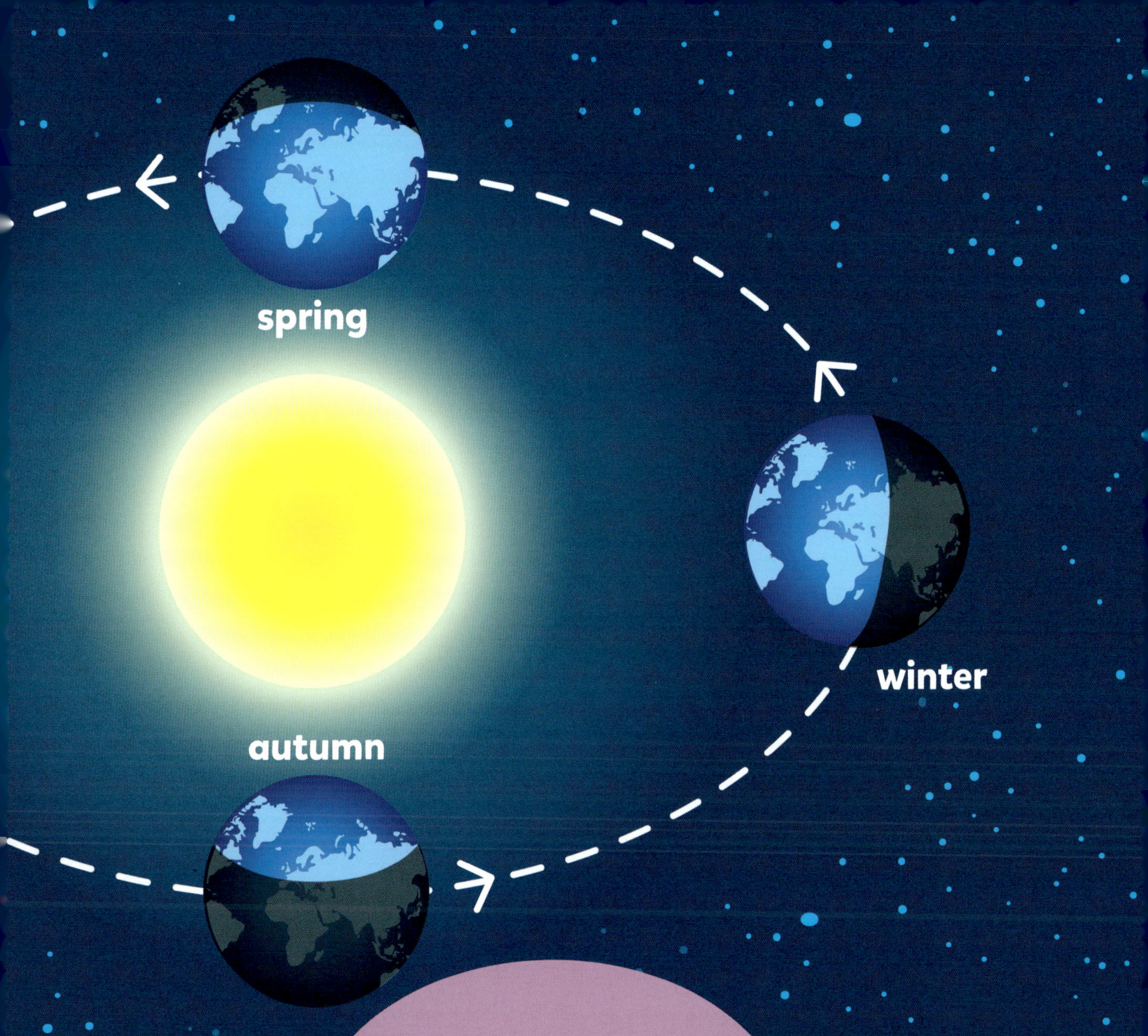

Why do you think Earth turns?

During the day, you are facing the sun.

At night, Earth turns, and you are away from the sun.

Earth gets light and heat from the sun.

How do light and heat help us?

Earth is where people, plants, and animals live.

Earth has air for
people to breathe.
Earth has food
for them to eat.

Earth has water for people, plants, and animals.

Most of Earth is covered with water.

Earth is a special planet.
We like to live on Earth!

You Connect!

What is something you like about Earth?

What is your favorite animal on Earth?

What other planets do you know about?

STEM Snapshot

Encourage students to think and ask questions like scientists. Ask the reader:

What is something you learned about Earth?

What is something you noticed about the sun?

What is something you still want to learn about Earth?

Photo Glossary

Earth

planets

solar system

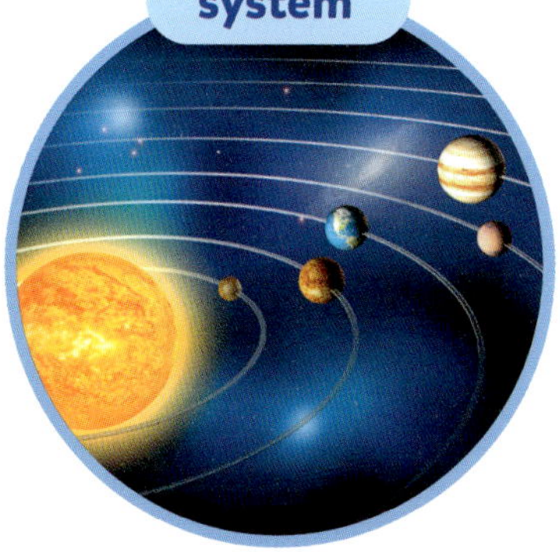

sun

Learn More

Armentrout, Patricia. *Our Amazing Earth*. New York: Crabtree, 2022.

Bauer, Marion Dane. *Earth*. New York: Simon Spotlight, 2021.

McAnulty, Stacy. *Earth! My First 4.54 Billion Years*. New York: Henry Holt, 2017.

Index

Photo Acknowledgments

The images in this book are used with the permission of: © andresr/iStockphoto, p. 11; © buradaki/iStockphoto, pp. 4–5, 16–17, 23 (Earth); © gradyreese/iStockphoto, p. 10; © KajaNi/iStockphoto, pp. 8–9; © Leopardinatree/iStockphoto, p. 18; © rwarnick/iStockphoto, pp. 6–7, 23 (planets, solar system, sun); © S847/iStockphoto, p. 16; © VR_Studio/iStockphoto, pp. 12–13; © vuk8691/iStockphoto, p. 19; © Wavebreakmedia/iStockphoto, p. 20; © YinYang/iStockphoto, pp. 14–15.

Cover Photo: © Paitoon Pornsuksomboon/Shutterstock Images.

Design Elements: © Mighty Media, Inc.

Lerner Publications Company
An imprint of Lerner Publishing Group, Inc.
241 First Avenue North
Minneapolis, MN 55401 USA

For reading levels and more information, look up this title at www.lernerbooks.com.

Main body text set in Mikado a Medium.
Typeface provided by Hannes von Doehren.

Library of Congress Cataloging-in-Publication Data

Names: Leed, Percy, 1968- author.
Title: Earth : a first look / Percy Leed.
Description: Minneapolis, MN : Lerner Publications, [2023] | Series: Read for a better world | Includes bibliographical references and index. | Audience: Ages 5-8 | Audience: Grades K–1 | Summary: "Earth is a special planet that has air, food, and water for people, plants, and animals. Easy-to-read text and fascinating photos give readers a first look at our home planet"– Provided by publisher.
Identifiers: LCCN 2021051327 | ISBN 9781728459219 (library binding) | ISBN 9781728464329 (paperback) | ISBN 9781728462172 (ebook)
Subjects: LCSH: Earth (Planet)—Juvenile literature.
Classification: LCC QB631.4 .L435 2023 | DDC 525–dc23/eng/20211216

LC record available at https://lccn.loc.gov/2021051327

Manufactured in the United States of America
1 - CG - 7/15/22